A Concise

Artificial Fly Fishing for Trout

Grey Drake

Alpha Editions

This edition published in 2021

ISBN : 9789355899316

Design and Setting By
Alpha Editions
www.alphaedis.com
Email – info@alphaedis.com

PREFACE.

IN the following humble effort I have endeavoured to communicate to the inexperienced lovers of artificial fly-fishing, as concisely as possible, and in a practical form, the result of my own experience of upwards of fifty years.

I treat only of fishing for trout with the artificial fly, adding a few observations on dibbing for trout with artificial flies and other baits.

I have no other ambition than that of initiating the tyro in the "gentle" and elegant art, with as little trouble and expense to him as may be. If he will do me the honour to become my disciple, and practise what I preach, I confidently promise him as much success as any artificial fly-fisher may reasonably expect.

GREY DRAKE.
1860.

CHAPTER I.

THE TACKLE.

I GIVE no directions for making rods, lines, or flies. I recommend the purchase of these at the best fishing-tackle shops. As to the supposed advantage to the artificial fly-fisher of being able to make artificial flies by the river-side, in imitation of the fly actually on the water, I am confidently of opinion, the acquisition of that art is wholly unnecessary and useless, as I shall more fully explain hereafter.

THE ROD.

The rod can scarcely be too light and pliable. Its pliability assists greatly, not only in throwing the fly, but in hooking and retaining the fish. The butt end should have a hollow sufficient to hold an extra top-piece, secured by a brass screw-nut, which, when the rod is used, should be taken out, and a spike screwed into its place. The spike is very useful for sticking the rod upright in the ground, as occasion may require. Some prefer a two-handed rod for large rivers. I think it is unnecessarily fatiguing to use a two-handed rod, inasmuch as a skilful artist can throw a fly with a single-handed rod as far as is necessary, and I never use any other. The single-handed rod should be about thirteen or fourteen feet long. To prevent the danger of breaking the rod, by the joints separating in throwing the fly, the joints should be whipped with strong silk, as shown in this figure. Fly rods are sometimes made to attach their joints by screws at the butt ends. Whipping rods so made is of course unnecessary, but I do not approve of screw-jointed rods, as they cannot be made sufficiently light and pliable.

THE REEL.

I prefer the spring cog-check wheel, which, when the line is lengthened or shortened, makes a noise like that caused by winding up a clock. By this reel the line may be lengthened or shortened with sufficient rapidity, and with precisely, and no more than the proper resistance, the checks preventing the line running out too fast. This reel is, I think, far preferable to the multiplying reel, which is very liable to get out of order.

THE LINE.

I prefer a horsehair line. In length it should be proportioned to the size of the river you fish. For large rivers seventy or eighty yards are not too much; for narrow rivers thirty or forty yards are sufficient. It should gradually taper towards the end to which the gut or tail line is attached, so that from four to five yards should be little thicker than the gut itself. The gut, or tail line, should be at least three yards long; thick and strong for rainy and windy weather and discoloured water. It cannot be too fine for bright weather and clear water, with little wind.

CHAPTER II.

ARTIFICIAL FLIES.

Volumes have been written on this subject, teaching the manner of making hundreds of different artificial flies, the materials and paraphernalia requisite for the finished fly-maker, the particular flies proper for various rivers, and for each month of the fishing season, &c., &c. All this, I consider, is perfectly useless, and I am decidedly of opinion, that when trout are disposed to take the fly, it matters not what fly is used, as to shape or colour, provided it be of the proper *size*. When trout are not disposed to take the fly, you may try all the flies in your book, without success. I have, by way of experiment, fished during an entire season with the coachman and governor only, and have been uniformly successful with those two flies, even during the May fly season, when the water has been covered with May flies, and the fish taking them greedily. The May fly is doubtless a great favourite with trout, and I would not recommend fishing with any other fly during the May fly season, although trout will take them before, as well as after the season. These flies make their appearance about the end of May, and disappear about the end of June.

Experience has taught me the fallacy of the common notion, that trout are finished entomologists, and will reject all flies not actually on the water, and even all flies in imitation of those actually on the water, unless the shape and size be exact, and the colour correct to a shade! The fact is, that when in the humour to take the fly, trout will take freely all sorts of insects that come in their way, from the May bug and grasshopper to the black gnat, and when feeding on insects they are not nice as to the kind, shape or colour of the insect presented to them. At the commencement of my piscatory career I was as fastidious as I imagined the fish to be, and I so continued until experience convinced me of my error.

I once met with a clergyman fishing, who cast his fly clumsily. He kept pretty well out of sight of the fish, but when his fly reached the water a large portion of the line accompanied it, making a splash, and frightening away the fish. He had a book full of all sorts of well-made[Pg 8] flies, which he constantly changed, but got no rises. A labouring countryman was following him at the distance of about a quarter of a mile, fishing with clumsy tackle, with which, by the skilful casting of his fly, he repeatedly took good-sized fish, to the great astonishment of the clergyman, who attributed his own want of success, not to his want of skill, but to his not using the right fly. "Will you permit me," said he to the countryman, "to look at your fly?"—"By all means," said he. "I am just about to put on another;" and, taking out his knife, he cut off a small piece of his black velveteen jacket, and stuck it on his hook, thus making what he called a black hackle! With that rude imitation he had caught all his fish; thus demonstrating that skill in the use of the artificial fly, however rudely made, will succeed, where the best imitations, clumsily used, will fail.

Although, by way of experiment, I have fished, during an entire season, with coachman and governor only, I would by no means recommend the fly-fisher to restrict himself to those flies; but I am quite sure that the flies comprised in the following list will be found amply sufficient for the whole fishing season, and for all countries and all rivers:—

The March Brown.	The Red Hackle.
The Governor.	The Black Hackle.
The Coachman.	The Blue Dun.
The Green Drake } May	The Alder Fly.
The Grey Drake } Flies.	The Black Gnat.

Fill your book with a sufficient quantity of these flies only, well made, half large and half small, and you will have as good a chance of success as a fly-fisher may reasonably expect.

The coachman is made with large peacock body, and white wings, and derives its name from the fact of its having been invented and first brought into notice by a Coachman, a celebrated fly-fisher. It is a very useful fly, and is taken by trout readily, in all waters, and in every part of the season, although not made to resemble any natural fly. It is preferable I think to the white moth for evening fishing.

Choose your May flies with wings made large and standing up, full bodies and long tails; and use no flies that are not made on Limerick hooks, which double your chance of hooking fish.

In all fly-fishing matches with which I am acquainted, and in some of which I have been myself engaged, each competitor has fished with a different kind of fly, and neither with a fly resembling that actually on the water. The success of each has been, generally, nearly equal, the winner gaining the match by a very few; attributable (as I believe), not to the fly he used, but to his superior skill, or to fortuitous circumstances, altogether independent of the particular fly he fished with.

I have dwelt upon this subject because I wish to guard the tyro against the too common failing of being fidgety as to his flies, and changing them repeatedly, fancying (for it is fancy only) that he does not get rises because he is not using the right fly. Fish with any of the flies I have mentioned—small, with fine gut, when the weather is bright and the water clear, with little wind—and larger, with stouter gut, when the weather is cloudy, windy, or rainy, and the water discoloured—and you may rest assured you will take as many fish as any competitor of only equal skill, with a book full of all sorts of flies, of all shapes and colours; and even with flies, admirably made by himself at the river-side, in imitation of the fly actually on the water.

I cannot, I think, better conclude my observations on artificial flies, or better satisfy the tyro of their truth, than by assuring him of the fact that some of the most successful first-rate fly-fishers, *old hands*, never, throughout the season, use any other flies than the red, brown, and black hackles, with and without wings, and the black gnat. I therefore hope that my list of flies will be considered amply sufficient, as I am quite sure *experience will prove it to be.*

CHAPTER III.

THROWING THE FLY.

Fly-fishing demands more skill than any other mode. To throw the fly well is the chief mystery of the art. Practice at first with a line only as long again as the rod, and lengthen it by degrees, as you find you progress. In drawing the line out of the water, incline the rod rather to the right, then describe a sort of half circle round your head by elevating the rod and giving it a motion towards your left. The moment the line thrown reaches its whole length, the fly should touch the water, else it will be checked, and recoil, falling heavily, and making a splash, which must by all means be avoided, or it will frighten away all the fish in the immediate neighborhood. Before you bring the rod forward to deliver the line, it (the line) should be at its full length *behind* you; if not, a splash will most infallibly ensue. As little as possible of the line should touch the water; to accomplish which, when you deliver the fly, the rod should not be depressed too much, for the nearer it approaches the water, the more line will fall upon it, and the greater splash will be the result.

As soon as the fly touches the water, draw it gently backwards, communicating to it an irregular motion by means of a tremulous movement of the wrist, causing it to imitate the movements of a fly accidentally cast on the water, and struggling to prevent drowning. This, especially if there be but little ripple, greatly increases your chance of a rise.

In bringing back the fly after having thrown it out, let it not approach too near to you before you raise it from the stream for another cast; otherwise, with a long line, you will find yourself so embarrassed as not to be able to give the line a sufficient swing back round your head to throw it with precision the next time.

Excellence in throwing the fly consists in causing it to fall *lightly*, and over any spot you may desire. This can only be accomplished by practice, for with all the knowledge theory can instil, it requires practice before you can throw the fly either to the exact spot you intend, or so that the sharpest eye cannot detect where it fell when there is a moderate ripple curling the surface of the water.

I have been diffuse in my directions for throwing the fly because it is the chief mystery in the art of fly-fishing, and difficult to be acquired in perfection. I strongly recommend the tyro to take a few lessons in throwing the fly from some experienced and skilful "Brother of the Angle." A few such lessons will be found to be worth volumes of theory.

CHAPTER IV.

GENERAL DIRECTIONS, OBSERVATIONS, ETC.

The fly-fisher may have acquired perfection in the art of throwing the fly; he may fish with the finest gut and the smallest and most killing flies; but *unless he keep out of sight of the fish*, he may just as well stay at home—he will take no fish. If a splash in the water, caused by the clumsy falling of the line, frighten away the fish, the sight of the fisherman himself will send them all to their holds, to a distance of thirty or forty yards from him! Trout are very sharp-sighted, timid and wary; and whenever they chance to see the fisherman, no bait whatever will be sufficient to tempt them to take it, and the utmost skill and dexterity will be thrown away.

When you observe a trout rise at a fly, throw your fly about a foot above where you judge his head to lie, and a little to the left or right of him. If he does not rise at your first cast, throw again three or four times. He will not take your fly unless it be presented to him temptingly, and near to him. He will not quit his post for your fly if it be out of his feeding circuit; and a few casts may bring it into that desirable locality. Trout always lie with their heads looking up the stream, watching for what it may bring them; and when they are taking the fly readily, they swim within a few inches of the surface of the water; but they will not go out of their feeding circuit to take *any* fly.

The very instant you perceive a trout has taken your fly, strike him *at the same instant* by slightly elevating the wrist. This should be done with the utmost rapidity, or the fish will manage to reject the treacherous imitation that has deceived him, and you will not rise him again for hours afterwards. In fact, I have often seen a good-sized trout that had escaped after having been hooked, not only afterwards invariably refuse the artificial fly, but quit his lair and take to his shelter the moment he perceived the tail line fall on the water.

When you have hooked a fish, you must necessarily act as the nature of the place will allow. If embarrassed with bushes, &c., get him out as quickly as possible. You may chance to lose him in the endeavour, but if you have not space for playing him, what is to be done? If you are in a situation to be able to play him, do so, keeping him well in hand with your bent rod. Never check a trout strongly in *his first* run, if avoidable. If he should be approaching anything that would endanger your line, strive to *guide* him *gradually* from it, by gently inclining your rod in the direction you wish him to take, always keeping him, as I before observed, well in hand with your bent rod. Never pull *directly against him*; for, if you do, you will probably cause him to plunge and leap in such a manner as to endanger your tackle, or tear the hook from its hold in his mouth. Trout, like many reasoning animals, may be easily *guided*, but never *compelled*, if of good size and strength, until, by playing him, he has been made too weary and exhausted for further contention. A small fish may of course be landed at once, but a fish of good size and strength should be *played*, if possible, until he becomes so exhausted by his struggles as to offer a favourable opportunity for introducing him into the landing net. If you have space for playing the fish, and are unencumbered by bushes, &c., perseverance, patience, address, and *sang froid*, will generally enable you to secure the largest trout.

It is difficult to give directions where to find trout in a trout stream. I have found them in every part of the stream. Good-sized trout often lurk near the edge under the banks, especially in narrow streams. I always try there first. They also lie in the currents of the stream watching for their prey. If there be any impediment in the stream, such as a large stone, &c., which, by projecting above or near the surface of the water, causes an increased rippling, never miss such a spot, but throw just above the rippling, drawing the fly through it. Towards evening trout are roaming about more freely in every part of the stream. It is of frequent occurrence to see a trout sailing up and down near the edge of the stream for a determinate distance. He is then in search of food. Keep out of sight, and he will probably take your fly.

Where trout are moderately plentiful, fish every yard of water.

As a rule, *small and fine* is the fly-fisher's maxim. In clear, bright water it is almost useless to use any thick-bodied fly. The smallest and thinnest-bodied flies are preferable in clear, bright water, and the larger in thick water, or on a windy day. You may successfully use any of the flies I have enumerated: small for clear, bright water, and larger for thick water or a blustering day.

The weather has an extraordinary effect on fish: I mean on their disposition to feed. In an easterly wind trout will not rise freely; thunder-storms they abominate; and very boisterous winds are unfavourable, let them proceed from what quarter they may. *During* and *after gentle showers*, with not too much wind, is the time, *par excellence*, for beguiling trout. Avoid a very bright day, unless there is sufficient wind to cause a strong ripple; but even then few trout will be your reward on a very bright day. A dark day succeeding a light night is never to be missed if you wish to fill your basket, for trout are almost as timid in a bright moonlight night as during the day. In such nights they will not feed freely. Should the next day, therefore, prove gloomy, it will probably repay you for many disappointments. In cold weather, fish only in the middle of the day: in hot weather, morning and evening are to be preferred. The evening is, I think, better than the morning; probably because, as trout abstain in a great measure from feeding during the heat, they are more eager when they recommence; and as they generally feed freely during the night, they are less eager for food in the morning. An hour before the disappearance of twilight, and, unless the night be *very* dark, an hour afterwards, will afford the best sport, and the largest fish. I once met with a singular proof of this. I had been fishing at Colonel Hawker's, Long Parish, Hants, and the day being very hot and bright, and no wind, I had bad sport. The keeper assured me, that if I waited until dark, and then fished a certain piece of backwater he pointed out, I should take some fine fish. Seeing by the movement of this water a fish was upon the feed during the twilight, I cast my fly for him, but as soon as it reached the water he was off. The keeper told me I was too early, that the fish were large and wary, that I must wait until it was *dark*. I did so, and putting on a large grey drake, in less than an hour I took four fine trout, weighing upwards of three pounds each. Although it was conveniently dark, the fish could see my fly, but could not see me or the line, and I could perceive a rise by a sort of bright flash in the water.

Do not allow your shadow to pass over the water if you can avoid it. You will rarely take a trout soon afterwards at the place where your shadow has passed over the water.

If you happen to be on the stream on a day when you have little sport, by all means repair to the same water the next day if you have reason to believe the trout to be moderately plentiful, and you will probably find them feeding freely. "Trout affection not long fasts," as dear old Isaac would say.

Skill in fly-fishing is neutralised by anger and impatience. Patience and perseverance stand at the head of the angler's cardinal virtues. With good tackle and proper-sized flies, moderate skill, and a favourable day, the tyro will astonish the natives of the stream if he keeps out of their sight; and if a little experience be added to the above, he may perchance astonish himself.

However fine the weather, wear long boots, as nearly waterproof as may be.

Frequently examine your fly to ascertain if it be in good order. I have often hooked a good trout, which soon got away, and, upon examination of my fly, I found the barb of the hook gone.

Take care that you do not, by a too sudden jerk, when bringing the fly forward for a fresh cast, snap it off. This often happens to the tyro, and sometimes to old hands. A slight, sharp, snapping noise of the line, in bringing it forward for a cast, is a sure symptom of the loss of the fly.

Never fish without carrying a landing net with you, or having it carried for you. The largest fish are frequently lost for want of a landing net, especially when you fish with small flies. If possible, the fish should never see the landing net, nor the person who uses it. Never allow the landing net to be *poked at* a fish; and never touch the line, nor allow it to be touched, whilst you have a good-sized fish at the end of it.

Carry with you, when fishing, a disengaging instrument, which you may screw to the butt end of the handle of your landing net, when required. The instrument is very useful for disengaging your fly and line from weeds, bushes, &c. It is sold in most of the fishing-tackle shops, and is in the form below. The edge *b* is made sufficiently sharp to cut away weeds, bushes, &c.

Always carry with you a piece of India rubber. Draw the tail line through it before you use the line, in order to straighten it and prove its strength; and if there are any faults in it, the India rubber will find them out, which is far better than making the discovery by losing a good fish from the too easy breaking of your untried tail line.

Gut is apt to snap if very dry, and I recommend immersing it in water for ten or fifteen minutes before using it. The best method of preserving gut that I know of is to keep it in parchment, slightly steeped in best salad oil.

Always carry with you some strong silk and strong thread, and a piece of shoemaker's wax.

CHAPTER V.

TO MAKE A TAIL LINE.

In joining pieces of gut together to make a tail line, I think the following joining knot, called the "sheet bend," is the best, as the knot is the smallest and neatest that can be made, and the more the line is stretched, the tighter the knot becomes.

Make a loop with the left-hand end of the gut to be joined, (*a b* figure below), and hold it between the finger and thumb of the left hand. Then pass the end of the right hand gut to be joined through the loop and under it; then round and under the two legs of the loop; then over the *b* leg of the loop; then under itself and out over the *a* leg of the loop, as shown in the figure below. (See another mode of joining pieces of gut for tail lines in the observations on bob-flies, page <u>20</u>.)

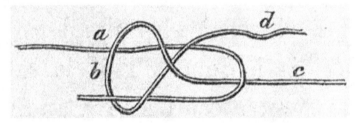

Although, perhaps, scarcely necessary, I may state that the hair line has a joining loop at the small end of it; the tail line a similar loop at each end, and the fly a similar loop at the other end of the gut to which it is attached. To attach the hair line to the tail line, insert the loop of the hair line, then bring the other end of the tail line through the loop of the hair line, and continue drawing it through that loop until both loops meet and interlace each other. The fly is attached to the tail line by interlacing the loops in the same way. By reversing the operation, the lines and fly may be readily detached and separated.

TO TIE ON A HOOK.

Take a sufficiency of strong silk, well waxed with shoemaker's wax. Flatten the gut to which you intend to fasten the hook; that is, about as much of it as, when the gut is placed on the hook, will reach half way down the shank. This may be done with the teeth or a pair of pliers, and is designed to prevent the gut from all chance of slipping. *Never omit this.* Lay the gut on the inside of the hook, and hold it between the thumb and finger of the left hand. Begin by wrapping the silk twice round the bare hook close to the end of the shank, then pass the silk over both gut and hook, winding it tightly on till you come near the bend; then fasten as follows:—When you come to within three turns of the distance to which you mean the silk to extend, lay the silk along the hook at *b* (figure below), leaving the end hanging down, take hold of the part of the silk *a*, and continue to wind it on in the same way, but making it pass over the silk at *b*, as well as over the gut and hook for three turns. Then take hold of the end of silk *c*, and pull all tight; cut off the end of silk *c*, and it is done. This is called the "invisible tie," and is the neatest I know, and the most secure.

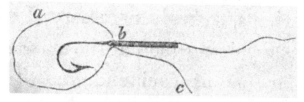

If you break a part of your rod, and have to splice it, fasten the splicing by the invisible tie. The splicing should be done with strong silk, well waxed with shoemaker's wax.

CHAPTER VI.

BOB-FLIES.

I never fish with more than one fly. Some fishermen use two, and even three, in addition to the end fly or stretcher. Those additional flies are called bob-flies. My opinion is, that one fly is sufficient, and that more are inconvenient, more difficult to manage, and cause a greater disturbance of the water, without any countervailing advantage. With bob-flies you may sometimes hook two fish at once, but they are generally very small under such circumstances. The luck of hooking two *good* fish at the same time rarely happens; and if it should happen, you would probably lose one or both of them, and some of your tackle into the bargain. Still, if you prefer fishing with two or more flies, the first bob-fly should be about three feet from the stretcher, and the second about five feet. More than two bob-flies I consider ridiculous, as well as prejudicial. The bob-fly may be attached, either by bending the tail line into a loop, thus

and putting on the bob-fly through the loop as you would put on the stretcher; or the tail line may be separated at the point where the bob-fly is to be attached, and the two ends of the gut, at the separation, may be placed

one over the other, thus .

Then tie a common knot at each end, thus

Then take the bob-fly, with about four inches of gut attached, and tie a common knot at the end of the gut. Then insert the bob-fly between the two ends of the gut and its knots. Then pull the two knots of the tail line tight together, and pull the knot of the bob-fly to meet those two knots, and the operation is finished, and the tail line and bob-fly assume the following appearance:—

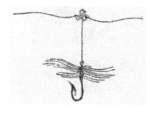

I consider this is the better plan of attaching a bob-fly, as it stands out better from the tail line, and is less likely to be entangled by it, and the knots of the tail line may be separated, and the bob-fly taken out. But, as I before observed, I do not recommend bob-flies. Those who like them may use them. Mrs. Glasse, after giving her admirable recipe for making a plum-pudding, adds, "those who like it may add an anchovy."

CHAPTER VII.

DIBBING.

Dibbing is another mode of fly-fishing, or rather of fishing on the surface; for other baits, besides the fly, may be used. It is more especially applicable to narrow streams that are embarrassed with trees and bushes, and is a most killing method.

In my fishing excursions, I always take with me a stiff little pocket-rod, of four pieces, each about two feet and a half in length, with a small reel attached, and about twenty yards of strong silk line, so that I may have a fair chance of successful sport, when I happen to be fishing on a stream where trees and bushes prevent my throwing the artificial fly, and where the large trout take refuge for the purpose of concealment, and the generally better supply of flies, insects, and other food. Your tackle must be very strong, for the larger trout only are generally taken in this way, and the trees and bushes give them a good chance, after being hooked, of escaping, by breaking your tackle.

In dibbing you can only use one fly. There should not be more than a couple of lengths of gut on the line. The gut must be strong, and so must that to which the fly is attached. Keep a few flies, tied to thicker gut than you use when throwing the fly.

Whenever you see a place between trees and bushes where a trout is likely to lie, drop the fly gently, *communicating to it a dancing movement.* The fly must only just touch the surface, the greatest care being taken that *not the smallest morsel of the gut touch the water.* This is most essential to success, for rarely indeed will you rise a trout by dibbing if he sees the least bit of gut in the stream.

It very frequently happens that you see a trout lying close to the edge of the stream, or under the shade of a bush. That fish, with care, you may be certain to rise. Never place yourself before the fish; but, standing behind him, drop the fly as directed, two or three inches on one side of his head, and not immediately before him. If you attempt to drop the fly *before* him, he will often see the gut, and vanish; whereas, by dropping it rather on one side, he is not aware of its approach until it touches the water. Thus he has no time to scrutinise too closely, for he will rise instantly, lest the fly pass away with the stream.

I have heard it asserted by very good fishermen, that dibbing ought not to be performed with the artificial fly, the deception being too obvious. I have however caught and seen caught many and fine fish by dibbing with the artificial fly.

In dibbing with the artificial fly, hackles are generally to be preferred. Any real fly that may be on the water, if of sufficient size to place on a hook, may be used in this mode of angling.

The flesh-fly will often kill; and the May bug and grasshopper are excellent baits. These should be thus baited:—Have double hooks, of various sizes, tied to a length of good strong gut. This gut must have a loop at one end to attach it to the other gut; which loop must be formed by *tying with silk, and not by means of a common knot*. To bait with this, insert the end of the loop or noose at the shoulder (directly at the back of the head) of the May bug, grasshopper or fly, and pass it through the body, bringing it out at the tail. Draw the insect along the gut till the shanks of the hooks *are buried in his body*, leaving only the points standing out on each side of the shoulders. The hooks should be of such a size as to extend a little beyond the bait.

Such I have found to be the neatest and best way of baiting with May bug, grasshopper, and flesh-fly, or other natural fly of sufficient size. The green and grey drakes, however, are too tender to be thus baited. A single hook must therefore be passed through the thickest part of the body, from side to side. The hook should not be very small, but have some of the shank broken off, for the shank should be short.

The gut for dibbing should not only be thick and strong, but should be died a palish blue, which may be thus easily done:—a wineglassful of common gin, having a teaspoonful of black ink mixed with it, must be made hot, and when rather cool, but by no means cold, steep the gut in it until it acquires the depth of colour you wish. The longer it remains in the mixture, the darker it becomes; but care must be taken that it be not *too* black.

For information on the subject of fishing trout rivers, streams, and lakes in the United Kingdom and France, I would refer the reader to the admirable work of Palmer Hackle, Esq., entitled "Hints on Angling" (Robinson, 69, Fleet-street), and to that of R. O'Connor, Esq., entitled "Field Sports of France" (John Murray, Albemarle-street). He will there find all the information he can desire, especially for fishing the numerous and well-stocked trout streams throughout France. Palmer Hackle agrees with me in the opinions I have expressed on the subject of artificial flies, and so does a French author, Mr. Guillemarde, who published a book on fishing in 1857 (Librairie de L. Hachette and Co., Rue Pierre-Sarazzen, No. 14, Paris). He observes (page 206) that five or six artificial flies of different sizes and colours are sufficient, and adds, "I know well that artificial fly-makers will not be of that opinion, and for a very good reason; but, independently of my own personal experience, I may refer to that of experienced professors, disinterested in the matter."

Very few Frenchmen are artificial fly-fishers. Mr. Guillemarde advises his countrymen to emulate the English, whom he compliments as *masters* in the elegant art, which, he says, they practise almost exclusively. He terms them "admirable fishermen," but spoils the compliment by assuring his readers, "the gentlemen of Great Britain fish in white cravats and kid gloves!"

First-rate fly-fishing may be had in all parts of France, and especially in the department of the Pas de Calais, in which are numerous admirable trout streams well stocked with fish, and where the sport may be enjoyed without interruption. Palmer Hackle, in his work, observes, "An angler who loves his art as none but anglers can, and desires to pursue his cherished recreation undisturbed by the malign influence of game-preservers, and unembittered by the sneers of money-getting fools, must visit the Continent. There he may roam unmolested and uncriticised if his deportment be that of a quiet, sensible man and a gentleman; and his sport will be such as to satisfy the most sanguine professor."

This, experience enables me fully to confirm, and it seems to be borne out by Mr. Guillemarde, who writes: "I speak of artificial fly-fishing, the most difficult but the most elegant mode of fishing with the line, and in which the preparation and execution, and the address of the professor, are most strikingly displayed and exemplified. Artificial fly-fishing is, at present, but little appreciated, or rather but little known, in France. Every year amateurs from England gather from our streams abundant harvests. It is a spectacle at once curious and humiliating to our national 'amour propre,' to observe the astonishment of most of our river-side inhabitants, endeavouring in vain to comprehend by what magic art these 'honourable gentlemen,' by flogging the air with their long switches, manage so easily to fill their baskets. May this little book contribute to popularise in our country those methods which are at present practised by so few, and which are equally agreeable and successful."

FINIS.

CPSIA information can be obtained
at www.ICGtesting.com
Printed in the USA
BVHW030455150222
628967BV00014B/224